Small Flock & Game BirD Log Book

Publish[illegible] Broh[illegible]

En[illegible]es, [illegible]

Grand [illegible]n, C[illegible]

[illegible]21

[illegible]N: 9[illegible]955[illegible]-5

Content

Bird Records - Space to record details on over 400 birds. page 1

Health Records - Log treatments of ill or injured birds. page 11

Feed Records - Keep a record of feed types, brands and quantities. page 14

Cage/Coop Record - Notes about the species/breed of birds in a particular cage or coop. Also know when you last did a general or deep cleaning of a cage. page 19

Egg Production Log - The log pages are designed to track egg production by cage or coop. Each page has room to record numbers for 5 cages/coops. There are enough pages for 40 cages. Keep track of daily, monthly and quarterly egg production and egg to hen ratio. page 28

Annual Total Egg Production Log - Notes to track annual egg production totals. page 92

Incubator Log - Space to track 90 hatch statistics. page 93

Expense Records - Track expenses by category. page 96

Income Records - Track sources of income from birds [eggs, meat, live bird sales, feather, manure...]. page 104

Profit/Loss Statement - Pages for five years. See at a glance how well your enterprise is doing. page 128

Contact List - Keep the contact information of veterinarians, suppliers, and buyers. page 133

Calculation Worksheets - Worksheets and samples of vital calculations: feed cost, amount of feed per bird, daily feed cost per bird, roost/hen ratio, average number of days per egg, cost per egg, cost per meat bird and cost per live bird sold. page 136

BIRD RECORDS

ID	NAME	BREED	SEX	HATCH DATE	PURCHASE DATE	LAY DATE	COST	NOTES

[illegible]

ID	NAME	[illegible]	SE	[illegible]	[illegible]	Y DATE	COST	NOTES

BIRD RECORDS

ID	NAME	BREED	SEX	HATCH DATE	PURCHASE DATE	LAY DATE	COST	NOTES

BIRD RECORDS

ID	NAME	BREED	SEX	HATCH DATE	PURCHASE DATE	LAY DATE	COST	NOTES

BIRD RECORDS

ID	NAME	BREED	SEX	HATCH DATE	PURCHASE DATE	LAY DATE	COST	NOTES

ID	NAME	[illegible]	SE[illegible]	[illegible]	[illegible]	[illegible] DATE	COST	NOTES

BIRD RECORDS

ID	NAME	BREED	SEX	HATCH DATE	PURCHASE DATE	LAY DATE	COST	NOTES

BIRD RECORDS

ID	NAME	BREED	SEX	HATCH DATE	PURCHASE DATE	LAY DATE	COST	NOTES

BIRD RECORDS

ID	NAME	BREED	SEX	HATCH DATE	PURCHASE DATE	LAY DATE	COST	NOTES

[illegible]

ID	NAME	[illegible]	SE[illegible]	[illegible]	PU[illegible] [illegible]Y DATE	COST	NOTES

HEALTH RECORDS

DATE	BIRD	ILLNESS	TREATMENT	NOTES

HEALTH RECORDS

DATE	BIRD	ILLNESS	TREATMENT	NOTES

HEALTH RECORDS

DATE	BIRD	ILLNESS	TREATMENT	NOTES

FEED RECORD

DATE	FEED DESCRIPTION	BRAND	QUANTITY	NOTES

FEED RECORDS

DATE	FEED DESCRIPTION	BRAND	QUANTITY	NOTES

FEED RECORDS

DATE	FEED DESCRIPTION	BRAND	QUANTITY	NOTES

FEED RECORDS

DATE	FEED DESCRIPTION	BRAND	QUANTITY	NOTES

FEED RECORD

DATE	FEED DESCRIPTION	BRAND	QUANTITY	[illegible]	NOTES

CAGE/COOP RECORDS

CAGE # **YEAR:** **Sq Ft:** ❐ BREEDERS ❐ LAYERS ❐ MEAT

DATE	CLEANING TYPE	HENS	ROOSTERS	HATCH DATE	BREED/STRAIN	NOTES

CAGE/COOP RECORDS

CAGE # **YEAR:** **Sq Ft:** ❒ BREEDERS ❒ LAYERS ❒ MEAT

DATE	CLEANING TYPE	HENS	ROOSTERS	HATCH DATE	BREED/STRAIN	NOTES

CAGE/COOP RECORDS

CAGE # **YEAR:** **Sq Ft:** ❐ BREEDERS ❐ LAYERS ❐ MEAT

DATE	CLEANING TYPE	HENS	ROOSTERS	HATCH DATE	BREED/STRAIN	NOTES

CAGE [illegible] RE[illegible]

CAGE #   YE[illegible]   [illegible]REEDERS   ❒ LAYERS   ❒ MEAT

DATE	CLEANING TYPE	[illegible]	[illegible]STE[illegible]	[illegible]DATE	[illegible]TRAIN	NOTES

CAGE/COOP RECORDS

CAGE # **YEAR:** **Sq Ft:** ❐ BREEDERS ❐ LAYERS ❐ MEAT

DATE	CLEANING TYPE	HENS	ROOSTERS	HATCH DATE	BREED/STRAIN	NOTES

CAGE/COOP RECORDS

CAGE # **YEAR:** **Sq Ft:** ❐ BREEDERS ❐ LAYERS ❐ MEAT

DATE	CLEANING TYPE	HENS	ROOSTERS	HATCH DATE	BREED/STRAIN	NOTES

CAGE/COOP RECORDS

CAGE # **YEAR:** **Sq Ft:** ❒ BREEDERS ❒ LAYERS ❒ MEAT

DATE	CLEANING TYPE	HENS	ROOSTERS	HATCH DATE	BREED/STRAIN	NOTES

CAGE [illegible] RE[illegible]

CAGE # YE[illegible] [illegible]REEDERS ❒ LAYERS ❒ MEAT

DATE	CLEANING TYPE	[illegible]	[illegible]ISTE[illegible]	[illegible] DATE	[illegible]TRAIN	NOTES

CAGE/COOP RECORDS

CAGE # **YEAR:** **Sq Ft:** ❒ BREEDERS ❒ LAYERS ❒ MEAT

DATE	CLEANING TYPE	HENS	ROOSTERS	HATCH DATE	BREED/STRAIN	NOTES

YEAR:
LOG #

EGG PRODUCTION RECORD

	CAGE #	1	2	3	4	5	6	7	8	9	10	11	12	13	14	15	16
		H/E	H/E	H/E	H/E	H/E	H/E	H/E	H/E	H/E	H/E	H/E	H/E	H/E	H/E	H/E	H/E
January																	
February																	
March																	
# OF HENS																	
DAILY EGG TOTAL																	

YEAR:
LOG #

EGG PRODUCTION RECORD

	CAGE #	17	18	19	20	21	22	23	24	25	26	27	28	29	30	31	Total Eggs
		H/E	H/E	H/E	H/E	H/E	H/E	H/E	H/E	H/E	H/E	H/E	H/E	H/E	H/E	H/E	
January																	
February																	
March																	
# OF HENS																	
DAILY EGG TOTAL																	

Egg Daily Avg:

Avg. # of Hens:

TOTAL EGGS FOR 1st QTR:

YEAR:

LOG #

EG D RECORD

CAGE #		1	2			5	6		8			11	12	13	14	15	16
		H/E	H/E			E	H		H/E			H/E	H/E	H/E	H/E	H/E	H/E
April																	
May																	
June																	
# OF HENS																	
DAILY EGG TOTAL																	

EGG PRODUCTION RECORD

YEAR:
LOG #

	CAGE #	17 H/E	18 H/E	19 H/E	20 H/E	21 H/E	22 H/E	23 H/E	24 H/E	25 H/E	26 H/E	27 H/E	28 H/E	29 H/E	30 H/E	31 H/E	Total Eggs
April																	
May																	
June																	
# OF HENS																	
DAILY EGG TOTAL																	

Egg Daily Avg:

Avg. # of Hens:

TOTAL EGGS FOR 2nd QTR:

YEAR:
LOG #

EGG PRODUCTION RECORD

	CAGE #	1	2	3	4	5	6	7	8	9	10	11	12	13	14	15	16
		H/E	H/E	H/E	H/E	H/E	H/E	H/E	H/E	H/E	H/E	H/E	H/E	H/E	H/E	H/E	H/E
July																	
August																	
September																	
# OF HENS																	
DAILY EGG TOTAL																	

YEAR:
LOG #

EGG PRODUCTION RECORD

	CAGE #	17 H/E	18 H/E	19 H/E	20 H/E	21 H/E	22 H/E	23 H/E	24 H/E	25 H/E	26 H/E	27 H/E	28 H/E	29 H/E	30 H/E	31 H/E	Total Eggs
July																	
August																	
September																	
# OF HENS																	
DAILY EGG TOTAL																	

Egg Daily Avg:

Avg # of Hens:

TOTAL EGGS FOR 3rd QTR:

YEAR:
LOG #

EG D RECORD

CAGE #	1	2									11	12	13	14	15	16
	H/E	H/E									H/E	H/E	H/E	H/E	H/E	H/E
October																
November																
December																
# OF HENS																
DAILY EGG TOTAL																

YEAR: LOG #	EGG PRODUCTION RECORD

	CAGE #	17	18	19	20	21	22	23	24	25	26	27	28	29	30	31	Total Eggs
		H/E	H/E	H/E	H/E	H/E	H/E	H/E	H/E	H/E	H/E	H/E	H/E	H/E	H/E	H/E	
October																	
November																	
December																	
# OF HENS																	
DAILY EGG TOTAL																	

Egg Daily Avg:	Avg. # of Hens:

TOTAL EGGS 1ST QTR	TOTAL EGGS 2ND QTR	TOTAL EGGS 3RD QTR	TOTAL EGGS 4TH QTR	TOTAL EGGS FOR YEAR:

YEAR:
LOG #

EGG PRODUCTION RECORD

	CAGE #	1	2	3	4	5	6	7	8	9	10	11	12	13	14	15	16
		H/E	H/E	H/E	H/E	H/E	H/E	H/E	H/E	H/E	H/E	H/E	H/E	H/E	H/E	H/E	H/E
January																	
February																	
March																	
# OF HENS																	
DAILY EGG TOTAL																	

YEAR:
LOG #

EGG PRODUCTION RECORD

	CAGE #	17	18	19	20	21	22	23	24	25	26	27	28	29	30	31	Total Eggs
		H/E	H/E	H/E	H/E	H/E	H/E	H/E	H/E	H/E	H/E	H/E	H/E	H/E	H/E	H/E	
January																	
February																	
March																	
# OF HENS																	
DAILY EGG TOTAL																	

Egg Daily Avg: | Avg. # of Hens:

TOTAL EGGS FOR 1st QTR:

YEAR:

LOG #

[illegible] RECORD

CAGE #		1	2	3	4	5	6	7	8	9	10	11	12	13	14	15	16
		H/E	H/E	H/E	H/E	H/E	H/E	H/E	H/E	H/E	H/E	H/E	H/E	H/E	H/E	H/E	H/E
April																	
May																	
June																	
# OF HENS																	
DAILY EGG TOTAL																	

YEAR:
LOG #

EGG PRODUCTION RECORD

	CAGE #	17	18	19	20	21	22	23	24	25	26	27	28	29	30	31	Total Eggs
		H/E	H/E	H/E	H/E	H/E	H/E	H/E	H/E	H/E	H/E	H/E	H/E	H/E	H/E	H/E	
April																	
May																	
June																	
# OF HENS																	
DAILY EGG TOTAL																	

Egg Daily Avg: | Avg. # of Hens:

TOTAL EGGS FOR 2nd QTR:

YEAR:
LOG #

EGG PRODUCTION RECORD

	CAGE #	1	2	3	4	5	6	7	8	9	10	11	12	13	14	15	16
		H/E	H/E	H/E	H/E	H/E	H/E	H/E	H/E	H/E	H/E	H/E	H/E	H/E	H/E	H/E	H/E
July																	
August																	
September																	
# OF HENS																	
DAILY EGG TOTAL																	

YEAR:
LOG #

EGG PRODUCTION RECORD

	CAGE #	17 H/E	18 H/E	19 H/E	20 H/E	21 H/E	22 H/E	23 H/E	24 H/E	25 H/E	26 H/E	27 H/E	28 H/E	29 H/E	30 H/E	31 H/E	Total Eggs
July																	
August																	
September																	
# OF HENS																	
DAILY EGG TOTAL																	

Egg Daily Avg:
Avg. # of Hens:

TOTAL EGGS FOR 3rd QTR:

YEAR:
LOG #

EGG PRODUCTION RECORD

CAGE #	1	2			5	6		8			11	12	13	14	15	16
	H/E	H/E			H/E			H/E			H/E	H/E	H/E	H/E	H/E	H/E
October																
November																
December																
# OF HENS																
DAILY EGG TOTAL																

YEAR:
LOG #

EGG PRODUCTION RECORD

	CAGE #	17 H/E	18 H/E	19 H/E	20 H/E	21 H/E	22 H/E	23 H/E	24 H/E	25 H/E	26 H/E	27 H/E	28 H/E	29 H/E	30 H/E	31 H/E	Total Eggs
October																	
November																	
December																	
# OF HENS																	
DAILY EGG TOTAL																	

Egg Daily Avg: Avg. # of Hens:

TOTAL EGGS 1ST QTR	TOTAL EGGS 2ND QTR	TOTAL EGGS 3RD QTR	TOTAL EGGS 4TH QTR	TOTAL EGGS FOR YEAR:

YEAR:
LOG #

EGG PRODUCTION RECORD

	CAGE #	1	2	3	4	5	6	7	8	9	10	11	12	13	14	15	16
		H/E	H/E	H/E	H/E	H/E	H/E	H/E	H/E	H/E	H/E	H/E	H/E	H/E	H/E	H/E	H/E
January																	
February																	
March																	
# OF HENS																	
DAILY EGG TOTAL																	

YEAR:
LOG #

EGG PRODUCTION RECORD

	CAGE #	17 H/E	18 H/E	19 H/E	20 H/E	21 H/E	22 H/E	23 H/E	24 H/E	25 H/E	26 H/E	27 H/E	28 H/E	29 H/E	30 H/E	31 H/E	Total Eggs
January																	
February																	
March																	
# OF HENS																	
DAILY EGG TOTAL																	

Egg Daily Avg: | Avg # of Hens:

TOTAL EGGS FOR 1st QTR:

YEAR:
LOG #

CAGE EGG PRODUCTION RECORD

	CAGE #	1	2	3	4	5	6	7	8	9	10	11	12	13	14	15	16
		H/E	H/E						H/E			H/E	H/E	H/E	H/E	H/E	H/E
April																	
May																	
June																	
# OF HENS																	
DAILY EGG TOTAL																	

YEAR:
LOG #

EGG PRODUCTION RECORD

	CAGE #	17	18	19	20	21	22	23	24	25	26	27	28	29	30	31	Total Eggs
		H/E	H/E	H/E	H/E	H/E	H/E	H/E	H/E	H/E	H/E	H/E	H/E	H/E	H/E	H/E	
April																	
May																	
June																	
# OF HENS																	
DAILY EGG TOTAL																	

Egg Daily Avg:

Avg. # of Hens:

TOTAL EGGS FOR 2nd QTR:

YEAR:
LOG #

EGG PRODUCTION RECORD

	CAGE #	1	2	3	4	5	6	7	8	9	10	11	12	13	14	15	16
		H/E	H/E	H/E	H/E	H/E	H/E	H/E	H/E	H/E	H/E	H/E	H/E	H/E	H/E	H/E	H/E
July																	
August																	
September																	
# OF HENS																	
DAILY EGG TOTAL																	

YEAR:
LOG #

EGG PRODUCTION RECORD

	CAGE #	17	18	19	20	21	22	23	24	25	26	27	28	29	30	31	Total Eggs
		H/E	H/E	H/E	H/E	H/E	H/E	H/E	H/E	H/E	H/E	H/E	H/E	H/E	H/E	H/E	
July																	
August																	
September																	
# OF HENS																	
DAILY EGG TOTAL																	

Egg Daily Avg:

Avg. # of Hens:

TOTAL EGGS FOR 3rd QTR:

YEAR:
LOG #

EG[illegible] D[illegible] RECORD

	CAGE #	1	2			5	6		8			11	12	13	14	15	16
		H/E	H/E			H/E	H/E		H/E			H/E	H/E	H/E	H/E	H/E	H/E
October																	
November																	
December																	
# OF HENS																	
DAILY EGG TOTAL																	

YEAR:
LOG #

EGG PRODUCTION RECORD

	CAGE #	17 H/E	18 H/E	19 H/E	20 H/E	21 H/E	22 H/E	23 H/E	24 H/E	25 H/E	26 H/E	27 H/E	28 H/E	29 H/E	30 H/E	31 H/E	Total Eggs
October																	
November																	
December																	
# OF HENS																	
DAILY EGG TOTAL																	

Egg Daily Avg:

Avg. # of Hens:

TOTAL EGGS 1ST QTR	TOTAL EGGS 2ND QTR	TOTAL EGGS 3RD QTR	TOTAL EGGS 4TH QTR

TOTAL EGGS FOR YEAR:

YEAR:
LOG #

EGG PRODUCTION RECORD

	CAGE #	1	2	3	4	5	6	7	8	9	10	11	12	13	14	15	16
		H/E	H/E	H/E	H/E	H/E	H/E	H/E	H/E	H/E	H/E	H/E	H/E	H/E	H/E	H/E	H/E
January																	
February																	
March																	
# OF HENS																	
DAILY EGG TOTAL																	

YEAR: LOG #	EGG PRODUCTION RECORD																
	CAGE #	**17** H/E	**18** H/E	**19** H/E	**20** H/E	**21** H/E	**22** H/E	**23** H/E	**24** H/E	**25** H/E	**26** H/E	**27** H/E	**28** H/E	**29** H/E	**30** H/E	**31** H/E	**Total Eggs**
January																	
February																	
March																	
# OF HENS																	
DAILY EGG TOTAL																	

Egg Daily Avg:	Avg # of Hens:
	TOTAL EGGS FOR 1st QTR:

YEAR:
LOG #

EGG PRODUCTION RECORD

CAGE #	1	2	3	4	5	6	7	8	9	10	11	12	13	14	15	16
	H/E	H/E	H/E	H/E	H/E	H/E	H/E	H/E	H/E	H/E	H/E	H/E	H/E	H/E	H/E	H/E
April																
May																
June																
# OF HENS																
DAILY EGG TOTAL																

YEAR:
LOG #

EGG PRODUCTION RECORD

	CAGE #	17	18	19	20	21	22	23	24	25	26	27	28	29	30	31	Total Eggs
		H/E	H/E	H/E	H/E	H/E	H/E	H/E	H/E	H/E	H/E	H/E	H/E	H/E	H/E	H/E	
April																	
May																	
June																	
# OF HENS																	
DAILY EGG TOTAL																	

Egg Daily Avg:

Avg. # of Hens:

TOTAL EGGS FOR 2nd QTR:

YEAR:
LOG #

EGG PRODUCTION RECORD

	CAGE #	1	2	3	4	5	6	7	8	9	10	11	12	13	14	15	16
		H/E	H/E	H/E	H/E	H/E	H/E	H/E	H/E	H/E	H/E	H/E	H/E	H/E	H/E	H/E	H/E
July																	
August																	
September																	
# OF HENS																	
DAILY EGG TOTAL																	

YEAR:
LOG #

EGG PRODUCTION RECORD

	CAGE #	17	18	19	20	21	22	23	24	25	26	27	28	29	30	31	Total Eggs
		H/E	H/E	H/E	H/E	H/E	H/E	H/E	H/E	H/E	H/E	H/E	H/E	H/E	H/E	H/E	
July																	
August																	
September																	
# OF HENS																	
DAILY EGG TOTAL																	

Egg Daily Avg:

Avg. # of Hens:

TOTAL EGGS FOR 3rd QTR:

YEAR:
LOG #

EGG PRODUCTION RECORD

CAGE #		1	2			5	6		8			11	12	13	14	15	16
		H/E	H/E			H/E	H/E		H/E			H/E	H/E	H/E	H/E	H/E	H/E
October																	
November																	
December																	
# OF HENS																	
DAILY EGG TOTAL																	

YEAR:
LOG #

EGG PRODUCTION RECORD

	CAGE #	17 H/E	18 H/E	19 H/E	20 H/E	21 H/E	22 H/E	23 H/E	24 H/E	25 H/E	26 H/E	27 H/E	28 H/E	29 H/E	30 H/E	31 H/E	Total Eggs
October																	
November																	
December																	
# OF HENS																	
DAILY EGG TOTAL																	

Egg Daily Avg:

Avg. # of Hens:

TOTAL EGGS 1ST QTR	TOTAL EGGS 2ND QTR	TOTAL EGGS 3RD QTR	TOTAL EGGS 4TH QTR

TOTAL EGGS FOR YEAR:

YEAR:
LOG #

EGG PRODUCTION RECORD

	CAGE #	1	2	3	4	5	6	7	8	9	10	11	12	13	14	15	16
		H/E	H/E	H/E	H/E	H/E	H/E	H/E	H/E	H/E	H/E	H/E	H/E	H/E	H/E	H/E	H/E
January																	
February																	
March																	
# OF HENS																	
DAILY EGG TOTAL																	

YEAR:
LOG #

EGG PRODUCTION RECORD

	CAGE #	17	18	19	20	21	22	23	24	25	26	27	28	29	30	31	Total Eggs
		H/E	H/E	H/E	H/E	H/E	H/E	H/E	H/E	H/E	H/E	H/E	H/E	H/E	H/E	H/E	
January																	
February																	
March																	
# OF HENS																	
DAILY EGG TOTAL																	

Egg Daily Avg:

Avg. # of Hens:

TOTAL EGGS FOR 1st QTR:

YEAR:

LOG #

EGG PRODUCTION RECORD

CAGE #	1	2	3	4	5	6	7	8	9	10	11	12	13	14	15	16
	H/E	H/E	H/E	H/E	H/E	H/E	H/E	H/E	H/E	H/E	H/E	H/E	H/E	H/E	H/E	H/E
April																
May																
June																
# OF HENS																
DAILY EGG TOTAL																

YEAR:
LOG #

EGG PRODUCTION RECORD

	CAGE #	17 H/E	18 H/E	19 H/E	20 H/E	21 H/E	22 H/E	23 H/E	24 H/E	25 H/E	26 H/E	27 H/E	28 H/E	29 H/E	30 H/E	31 H/E	Total Eggs
April																	
May																	
June																	
# OF HENS																	
DAILY EGG TOTAL																	

Egg Daily Avg: | Avg. # of Hens:

TOTAL EGGS FOR 2nd QTR:

YEAR:
LOG #

EGG PRODUCTION RECORD

	CAGE #	1	2	3	4	5	6	7	8	9	10	11	12	13	14	15	16
		H/E	H/E	H/E	H/E	H/E	H/E	H/E	H/E	H/E	H/E	H/E	H/E	H/E	H/E	H/E	H/E
July																	
August																	
September																	
# OF HENS																	
DAILY EGG TOTAL																	

YEAR:
LOG #

EGG PRODUCTION RECORD

	CAGE #	17 H/E	18 H/E	19 H/E	20 H/E	21 H/E	22 H/E	23 H/E	24 H/E	25 H/E	26 H/E	27 H/E	28 H/E	29 H/E	30 H/E	31 H/E	Total Eggs
July																	
August																	
September																	
# OF HENS																	
DAILY EGG TOTAL																	

Egg Daily Avg:

Avg. # of Hens:

TOTAL EGGS FOR 3rd QTR:

EGG PRODUCTION RECORD

YEAR:

LOG #

	CAGE #	1	2				6		8			11	12	13	14	15	16
		H/E	H/E			H/E	H/E		H/E			H/E	H/E	H/E	H/E	H/E	H/E
October																	
November																	
December																	
# OF HENS																	
DAILY EGG TOTAL																	

YEAR:
LOG #

EGG PRODUCTION RECORD

	CAGE #	17 H/E	18 H/E	19 H/E	20 H/E	21 H/E	22 H/E	23 H/E	24 H/E	25 H/E	26 H/E	27 H/E	28 H/E	29 H/E	30 H/E	31 H/E	Total Eggs
October																	
November																	
December																	
# OF HENS																	
DAILY EGG TOTAL																	

Egg Daily Avg:

Avg. # of Hens:

TOTAL EGGS 1ST QTR	TOTAL EGGS 2ND QTR	TOTAL EGGS 3RD QTR	TOTAL EGGS 4TH QTR

TOTAL EGGS FOR YEAR:

YEAR:
LOG #

EGG PRODUCTION RECORD

	CAGE #	1	2	3	4	5	6	7	8	9	10	11	12	13	14	15	16
		H/E	H/E	H/E	H/E	H/E	H/E	H/E	H/E	H/E	H/E	H/E	H/E	H/E	H/E	H/E	H/E
January																	
February																	
March																	
# OF HENS																	
DAILY EGG TOTAL																	

YEAR: LOG #	EGG PRODUCTION RECORD															
	CAGE #	17 H/E	18 H/E	19 H/E	20 H/E	21 H/E	22 H/E	23 H/E	24 H/E	25 H/E	26 H/E	27 H/E	28 H/E	29 H/E	30 H/E	31 H/E	Total Eggs
January																	
February																	
March																	
# OF HENS																	
DAILY EGG TOTAL																	

Egg Daily Avg:

Avg # of Hens:

TOTAL EGGS FOR 1st QTR:

YEAR:
LOG #

EGG PRODUCTION RECORD

CAGE #		1	2	[illegible]	[illegible]	5	6	[illegible]	8	[illegible]	[illegible]	11	12	13	14	15	16
		H/E	H/E			H/E	H/E		H/E			H/E	H/E	H/E	H/E	H/E	H/E
April																	
May																	
June																	
# OF HENS																	
DAILY EGG TOTAL																	

YEAR:
LOG #

EGG PRODUCTION RECORD

	CAGE #	17 H/E	18 H/E	19 H/E	20 H/E	21 H/E	22 H/E	23 H/E	24 H/E	25 H/E	26 H/E	27 H/E	28 H/E	29 H/E	30 H/E	31 H/E	Total Eggs
April																	
May																	
June																	
# OF HENS																	
DAILY EGG TOTAL																	

Egg Daily Avg: | Avg. # of Hens:

TOTAL EGGS FOR 2nd QTR:

YEAR:
LOG #

EGG PRODUCTION RECORD

CAGE #		1	2	3	4	5	6	7	8	9	10	11	12	13	14	15	16
		H/E	H/E	H/E	H/E	H/E	H/E	H/E	H/E	H/E	H/E	H/E	H/E	H/E	H/E	H/E	H/E
July																	
August																	
September																	
# OF HENS																	
DAILY EGG TOTAL																	

YEAR:
LOG #

EGG PRODUCTION RECORD

	CAGE #	17	18	19	20	21	22	23	24	25	26	27	28	29	30	31	Total Eggs
		H/E	H/E	H/E	H/E	H/E	H/E	H/E	H/E	H/E	H/E	H/E	H/E	H/E	H/E	H/E	
July																	
August																	
September																	
# OF HENS																	
DAILY EGG TOTAL																	

Egg Daily Avg: | Avg. # of Hens:

TOTAL EGGS FOR 3rd QTR:

YEAR:
LOG #

EG[illegible] RECORD

CAGE #	1	2									11	12	13	14	15	16
	H/E	H/E									H/E	H/E	H/E	H/E	H/E	H/E
October																
November																
December																
# OF HENS																
DAILY EGG TOTAL																

YEAR: LOG #	EGG PRODUCTION RECORD

	CAGE #	17 H/E	18 H/E	19 H/E	20 H/E	21 H/E	22 H/E	23 H/E	24 H/E	25 H/E	26 H/E	27 H/E	28 H/E	29 H/E	30 H/E	31 H/E	Total Eggs
October																	
November																	
December																	
# OF HENS																	
DAILY EGG TOTAL																	

Egg Daily Avg:	Avg. # of Hens:

TOTAL EGGS 1ST QTR	TOTAL EGGS 2ND QTR	TOTAL EGGS 3RD QTR	TOTAL EGGS 4TH QTR	**TOTAL EGGS FOR YEAR:**

YEAR:
LOG #

EGG PRODUCTION RECORD

	CAGE #	1	2	3	4	5	6	7	8	9	10	11	12	13	14	15	16
		H/E	H/E	H/E	H/E	H/E	H/E	H/E	H/E	H/E	H/E	H/E	H/E	H/E	H/E	H/E	H/E
January																	
February																	
March																	
# OF HENS																	
DAILY EGG TOTAL																	

YEAR:
LOG #

EGG PRODUCTION RECORD

	CAGE #	17	18	19	20	21	22	23	24	25	26	27	28	29	30	31	Total Eggs
		H/E	H/E	H/E	H/E	H/E	H/E	H/E	H/E	H/E	H/E	H/E	H/E	H/E	H/E	H/E	
January																	
February																	
March																	
# OF HENS																	
DAILY EGG TOTAL																	

Egg Daily Avg: | Avg. # of Hens:

TOTAL EGGS FOR 1st QTR:

YEAR:
LOG #

EGG PRODUCTION RECORD

CAGE #	1	2	3	4	5	6	7	8	9	10	11	12	13	14	15	16
	H/E	H/E	H/E	H/E	H/E	H/E	H/E	H/E	H/E	H/E	H/E	H/E	H/E	H/E	H/E	H/E
April																
May																
June																
# OF HENS																
DAILY EGG TOTAL																

YEAR: LOG #	EGG PRODUCTION RECORD

	CAGE #	17 H/E	18 H/E	19 H/E	20 H/E	21 H/E	22 H/E	23 H/E	24 H/E	25 H/E	26 H/E	27 H/E	28 H/E	29 H/E	30 H/E	31 H/E	Total Eggs
April																	
May																	
June																	
# OF HENS																	
DAILY EGG TOTAL																	

Egg Daily Avg:	Avg. # of Hens:

TOTAL EGGS FOR 2nd QTR:

YEAR:
LOG #

EGG PRODUCTION RECORD

CAGE #		1	2	3	4	5	6	7	8	9	10	11	12	13	14	15	16
		H/E	H/E	H/E	H/E	H/E	H/E	H/E	H/E	H/E	H/E	H/E	H/E	H/E	H/E	H/E	H/E
July																	
August																	
September																	
# OF HENS																	
DAILY EGG TOTAL																	

YEAR:
LOG #

EGG PRODUCTION RECORD

	CAGE #	17 H/E	18 H/E	19 H/E	20 H/E	21 H/E	22 H/E	23 H/E	24 H/E	25 H/E	26 H/E	27 H/E	28 H/E	29 H/E	30 H/E	31 H/E	Total Eggs
July																	
August																	
September																	
# OF HENS																	
DAILY EGG TOTAL																	

Egg Daily Avg:

Avg. # of Hens:

TOTAL EGGS FOR 3rd QTR:

YEAR:
LOG #

EGG PRODUCTION RECORD

CAGE #	1	2			5	6		8			11	12	13	14	15	16
	H/E	H/E	H/E	H/E	H/E	H/E	H/E	H/E	H/E	H/E	H/E	H/E	H/E	H/E	H/E	H/E
October																
November																
December																
# OF HENS																
DAILY EGG TOTAL																

YEAR:
LOG #

EGG PRODUCTION RECORD

	CAGE #	17 H/E	18 H/E	19 H/E	20 H/E	21 H/E	22 H/E	23 H/E	24 H/E	25 H/E	26 H/E	27 H/E	28 H/E	29 H/E	30 H/E	31 H/E	Total Eggs
October																	
November																	
December																	
# OF HENS																	
DAILY EGG TOTAL																	

Egg Daily Avg:
Avg. # of Hens:

TOTAL EGGS 1ST QTR	TOTAL EGGS 2ND QTR	TOTAL EGGS 3RD QTR	TOTAL EGGS 4TH QTR

TOTAL EGGS FOR YEAR:

YEAR:
LOG #

EGG PRODUCTION RECORD

	CAGE #	1	2	3	4	5	6	7	8	9	10	11	12	13	14	15	16
		H/E	H/E	H/E	H/E	H/E	H/E	H/E	H/E	H/E	H/E	H/E	H/E	H/E	H/E	H/E	H/E
January																	
February																	
March																	
# OF HENS																	
DAILY EGG TOTAL																	

YEAR:
LOG #

EGG PRODUCTION RECORD

	CAGE #	17	18	19	20	21	22	23	24	25	26	27	28	29	30	31	Total Eggs
		H/E	H/E	H/E	H/E	H/E	H/E	H/E	H/E	H/E	H/E	H/E	H/E	H/E	H/E	H/E	
January																	
February																	
March																	
# OF HENS																	
DAILY EGG TOTAL																	

Egg Daily Avg: | Avg # of Hens:

TOTAL EGGS FOR 1st QTR:

YEAR:
LOG #

EGG PRODUCTION RECORD

CAGE #	1	2			5	6		8			11	12	13	14	15	16
	H/E	H/E			H/E	H/E		H/E			H/E	H/E	H/E	H/E	H/E	H/E
April																
May																
June																
# OF HENS																
DAILY EGG TOTAL																

YEAR:
LOG #

EGG PRODUCTION RECORD

	CAGE #	17 H/E	18 H/E	19 H/E	20 H/E	21 H/E	22 H/E	23 H/E	24 H/E	25 H/E	26 H/E	27 H/E	28 H/E	29 H/E	30 H/E	31 H/E	Total Eggs
April																	
May																	
June																	
# OF HENS																	
DAILY EGG TOTAL																	

Egg Daily Avg: | Avg. # of Hens:

TOTAL EGGS FOR 2nd QTR:

YEAR:
LOG #

EGG PRODUCTION RECORD

CAGE #		1	2	3	4	5	6	7	8	9	10	11	12	13	14	15	16
		H/E	H/E	H/E	H/E	H/E	H/E	H/E	H/E	H/E	H/E	H/E	H/E	H/E	H/E	H/E	H/E
July																	
August																	
September																	
# OF HENS																	
DAILY EGG TOTAL																	

YEAR:
LOG #

EGG PRODUCTION RECORD

	CAGE #	17 H/E	18 H/E	19 H/E	20 H/E	21 H/E	22 H/E	23 H/E	24 H/E	25 H/E	26 H/E	27 H/E	28 H/E	29 H/E	30 H/E	31 H/E	Total Eggs
July																	
August																	
September																	
# OF HENS																	
DAILY EGG TOTAL																	

Egg Daily Avg:
Avg. # of Hens:

TOTAL EGGS FOR 3rd QTR:

YEAR:
LOG #

EGG PRODUCTION RECORD

CAGE #	1	2			5	6		8			11	12	13	14	15	16
	H/E	H/E			H/E	H/E		H/E			H/E	H/E	H/E	H/E	H/E	H/E
October																
November																
December																
# OF HENS																
DAILY EGG TOTAL																

YEAR:
LOG #

EGG PRODUCTION RECORD

	CAGE #	17 H/E	18 H/E	19 H/E	20 H/E	21 H/E	22 H/E	23 H/E	24 H/E	25 H/E	26 H/E	27 H/E	28 H/E	29 H/E	30 H/E	31 H/E	Total Eggs
October																	
November																	
December																	
# OF HENS																	
DAILY EGG TOTAL																	

Egg Daily Avg:

Avg. # of Hens:

TOTAL EGGS 1ST QTR	TOTAL EGGS 2ND QTR	TOTAL EGGS 3RD QTR	TOTAL EGGS 4TH QTR	TOTAL EGGS FOR YEAR:

ANNUAL EGG PRODUCTION TOTALS

YEAR	LOG #	DAILY AVG.	HEN # AVG.	EGGS PRODUCED	EGG/HEN AVG.	**TOTAL EGGS FOR YEAR**

INCUBATOR RECORDS

DATE	SPECIES/BREED	# EGGS STARTED	# INFERTILE	# HATCED	HATCH DATE	# UNHATCHED	# REACHING MATURITY	# OF HENS ADDED TO FLOCK	# OF HENS SOLD	# OF ROOSTERS ADDED TO FLOCK	# OF ROOSTERS SOLD	NOTES

INC[illegible] RE[illegible]

DATE	SPECIES/BREED	# EGGS S[illegible]	[illegible]	HATCH DA[illegible]	MATURITY [illegible]	[illegible] ROOSTERS ADDED TO FLOCK	# OF ROOSTERS SOLD	NOTES

INCUBATOR RECORDS

DATE	SPECIES/BREED	# EGGS STARTED	# INFERTILE	# HATCED	HATCH DATE	# UNHATCHED	# REACHING MATURITY	# OF HENS ADDED TO FLOCK	# OF HENS SOLD	# OF ROOSTERS ADDED TO FLOCK	# OF ROOSTERS SOLD	NOTES

EXPENDITURES

DATE	TO	DESCRIPTION	FEED	SUPPLIES	EQUIPMENT	MAINTENANCE	MEDICAL	OTHER	TOTAL

EXPENDITURES

DATE	TO	DESCRIPTION	FEED	SUPPLIES	EQUIPMENT	MAINTENANCE	MEDICAL	OTHER	TOTAL

EXPENDITURES

DATE	TO	DESCRIPTION		EQUIPMENT	MAINTENANCE	MEDICAL	OTHER	TOTAL

EXPENDITURES

DATE	TO	DESCRIPTION	FEED	SUPPLIES	EQUIPMENT	MAINTENANCE	MEDICAL	OTHER	TOTAL

EXPENDITURES

DATE	TO	DESCRIPTION	FEED	SUPPLIES	EQUIPMENT	MAINTENANCE	MEDICAL	OTHER	TOTAL

EXPENDITURES

DATE	TO	DESCRIPTION	FEED	SUPPLIES	EQUIPMENT	MAINTENANCE	MEDICAL	OTHER	TOTAL

EXPENDITURES

DATE	TO	DESCRIPTION	[illegible]	EQUIPMENT	MAINTENANCE	MEDICAL	OTHER	TOTAL

EXPENDITURES

DATE	TO	DESCRIPTION	FEED	SUPPLIES	EQUIPMENT	MAINTENANCE	MEDICAL	OTHER	TOTAL

INCOME RECORDS

	DATE	SOLD TO	CONTACT INFO	EGGS	
				QTY	$
1					
2					
3					
4					
5					
6					
7					
8					
9					
10					
11					
12					
13					
14					
15					
16					
17					
18					
19					
20					
21					
22					
23					
24					
25					
26					
27					
28					
29					
30					
31					

	CHICKS		BIRDS		MEAT		OTHER			
	QTY	$	QTY	$	QTY	$	ITEM	QTY	$	TOTAL
1										
2										
3										
4										
5										
6										
7										
8										
9										
10										
11										
12										
13										
14										
15										
16										
17										
18										
19										
20										
21										
22										
23										
24										
25										
26										
27										
28										
29										
30										
31										

	DATE	SC… …INFO	EGGS	
			QTY	$
1				
2				
3				
4				
5				
6				
7				
8				
9				
10				
11				
12				
13				
14				
15				
16				
17				
18				
19				
20				
21				
22				
23				
24				
25				
26				
27				
28				
29				
30				
31				

	CHICKS		BIRDS		MEAT		OTHER			TOTAL
	QTY	$	QTY	$	QTY	$	ITEM	QTY	$	
1										
2										
3										
4										
5										
6										
7										
8										
9										
10										
11										
12										
13										
14										
15										
16										
17										
18										
19										
20										
21										
22										
23										
24										
25										
26										
27										
28										
29										
30										
31										

INCOME RECORDS

	DATE	SOLD TO	CONTACT INFO	EGGS	
				QTY	$
1					
2					
3					
4					
5					
6					
7					
8					
9					
10					
11					
12					
13					
14					
15					
16					
17					
18					
19					
20					
21					
22					
23					
24					
25					
26					
27					
28					
29					
30					
31					

	CHICKS		BIRDS		MEAT		OTHER			
	QTY	$	QTY	$	QTY	$	ITEM	QTY	$	TOTAL
1										
2										
3										
4										
5										
6										
7										
8										
9										
10										
11										
12										
13										
14										
15										
16										
17										
18										
19										
20										
21										
22										
23										
24										
25										
26										
27										
28										
29										
30										
31										

OME RDS

	DATE	SC	INFO	EGGS QTY	EGGS $
1					
2					
3					
4					
5					
6					
7					
8					
9					
10					
11					
12					
13					
14					
15					
16					
17					
18					
19					
20					
21					
22					
23					
24					
25					
26					
27					
28					
29					
30					
31					

	CHICKS		BIRDS		MEAT		OTHER			
	QTY	$	QTY	$	QTY	$	ITEM	QTY	$	TOTAL
1										
2										
3										
4										
5										
6										
7										
8										
9										
10										
11										
12										
13										
14										
15										
16										
17										
18										
19										
20										
21										
22										
23										
24										
25										
26										
27										
28										
29										
30										
31										

INCOME RECORDS

	DATE	SOLD TO	CONTACT INFO	EGGS	
				QTY	$
1					
2					
3					
4					
5					
6					
7					
8					
9					
10					
11					
12					
13					
14					
15					
16					
17					
18					
19					
20					
21					
22					
23					
24					
25					
26					
27					
28					
29					
30					
31					

	CHICKS		BIRDS		MEAT		OTHER			
	QTY	$	QTY	$	QTY	$	ITEM	QTY	$	TOTAL
1										
2										
3										
4										
5										
6										
7										
8										
9										
10										
11										
12										
13										
14										
15										
16										
17										
18										
19										
20										
21										
22										
23										
24										
25										
26										
27										
28										
29										
30										
31										

OME RDS

	DATE	SC	INFO	EGGS	
				QTY	$
1					
2					
3					
4					
5					
6					
7					
8					
9					
10					
11					
12					
13					
14					
15					
16					
17					
18					
19					
20					
21					
22					
23					
24					
25					
26					
27					
28					
29					
30					
31					

	CHICKS		BIRDS		MEAT		OTHER			TOTAL
	QTY	$	QTY	$	QTY	$	ITEM	QTY	$	
1										
2										
3										
4										
5										
6										
7										
8										
9										
10										
11										
12										
13										
14										
15										
16										
17										
18										
19										
20										
21										
22										
23										
24										
25										
26										
27										
28										
29										
30										
31										

INCOME RECORDS

	DATE	SOLD TO	CONTACT INFO	EGGS	
				QTY	$
1					
2					
3					
4					
5					
6					
7					
8					
9					
10					
11					
12					
13					
14					
15					
16					
17					
18					
19					
20					
21					
22					
23					
24					
25					
26					
27					
28					
29					
30					
31					

	CHICKS		BIRDS		MEAT		OTHER			
	QTY	$	QTY	$	QTY	$	ITEM	QTY	$	TOTAL
1										
2										
3										
4										
5										
6										
7										
8										
9										
10										
11										
12										
13										
14										
15										
16										
17										
18										
19										
20										
21										
22										
23										
24										
25										
26										
27										
28										
29										
30										
31										

[illegible]OME[illegible]RDS

	DATE	SO[illegible]	[illegible] INFO	EGGS	
				QTY	$
1					
2					
3					
4					
5					
6					
7					
8					
9					
10					
11					
12					
13					
14					
15					
16					
17					
18					
19					
20					
21					
22					
23					
24					
25					
26					
27					
28					
29					
30					
31					

	CHICKS		BIRDS		MEAT		OTHER			
	QTY	$	QTY	$	QTY	$	ITEM	QTY	$	TOTAL
1										
2										
3										
4										
5										
6										
7										
8										
9										
10										
11										
12										
13										
14										
15										
16										
17										
18										
19										
20										
21										
22										
23										
24										
25										
26										
27										
28										
29										
30										
31										

INCOME RECORDS

	DATE	SOLD TO	CONTACT INFO	EGGS	
				QTY	$
1					
2					
3					
4					
5					
6					
7					
8					
9					
10					
11					
12					
13					
14					
15					
16					
17					
18					
19					
20					
21					
22					
23					
24					
25					
26					
27					
28					
29					
30					
31					

	CHICKS		BIRDS		MEAT		OTHER			
	QTY	$	QTY	$	QTY	$	ITEM	QTY	$	TOTAL
1										
2										
3										
4										
5										
6										
7										
8										
9										
10										
11										
12										
13										
14										
15										
16										
17										
18										
19										
20										
21										
22										
23										
24										
25										
26										
27										
28										
29										
30										
31										

OME RD

	DATE	SC	INFO	EGGS	
				QTY	$
1					
2					
3					
4					
5					
6					
7					
8					
9					
10					
11					
12					
13					
14					
15					
16					
17					
18					
19					
20					
21					
22					
23					
24					
25					
26					
27					
28					
29					
30					
31					

	CHICKS		BIRDS		MEAT		OTHER			
	QTY	$	QTY	$	QTY	$	ITEM	QTY	$	TOTAL
1										
2										
3										
4										
5										
6										
7										
8										
9										
10										
11										
12										
13										
14										
15										
16										
17										
18										
19										
20										
21										
22										
23										
24										
25										
26										
27										
28										
29										
30										
31										

INCOME RECORDS

	DATE	SOLD TO	CONTACT INFO	EGGS	
				QTY	$
1					
2					
3					
4					
5					
6					
7					
8					
9					
10					
11					
12					
13					
14					
15					
16					
17					
18					
19					
20					
21					
22					
23					
24					
25					
26					
27					
28					
29					
30					
31					

	CHICKS		BIRDS		MEAT		OTHER			TOTAL
	QTY	$	QTY	$	QTY	$	ITEM	QTY	$	
1										
2										
3										
4										
5										
6										
7										
8										
9										
10										
11										
12										
13										
14										
15										
16										
17										
18										
19										
20										
21										
22										
23										
24										
25										
26										
27										
28										
29										
30										
31										

OME RDS

	DATE	SC	INFO	EGGS	
				QTY	$
1					
2					
3					
4					
5					
6					
7					
8					
9					
10					
11					
12					
13					
14					
15					
16					
17					
18					
19					
20					
21					
22					
23					
24					
25					
26					
27					
28					
29					
30					
31					

	CHICKS		BIRDS		MEAT		OTHER			
	QTY	$	QTY	$	QTY	$	ITEM	QTY	$	TOTAL
1										
2										
3										
4										
5										
6										
7										
8										
9										
10										
11										
12										
13										
14										
15										
16										
17										
18										
19										
20										
21										
22										
23										
24										
25										
26										
27										
28										
29										
30										
31										

PROFIT/LOSS STATEMENT

for January 1, 20_____ to December 31, 20_____

EXPENDITURES

Feed	
Supplies	
Equipment	
Maintenance/Repair	
Medical/Vet	
Other:	
______________	
______________	
______________	
______________	
______________	

INCOME

Sales	
Other	

TOTAL EXPENDITURES ____________________

TOTAL INCOME ____________________

PROFIT / (LOSS) ____________________

PROFIT/LOSS STATEMENT

for January 1, 20_____ to December 31, 20_____

EXPENDITURES

Feed	
Supplies	
Equipment	
Maintenance/Repair	
Medical/Vet	
Other:	
_______________	
_______________	
_______________	
_______________	
_______________	

INCOME

Sales	
Other	

TOTAL EXPENDITURES _______________

TOTAL INCOME _______________

PROFIT / (LOSS) _______________

EXPENDITURES

- Feed
- Supplies
- Equipment
- Maintenance/Re
- Medical/Vet
- Other:

ICOME

Sales	
Other	

IRES ______________________

OME ______________________

PR SS) ______________________

PROFIT/LOSS STATEMENT

for January 1, 20_____ to December 31, 20_____

EXPENDITURES

Feed	
Supplies	
Equipment	
Maintenance/Repair	
Medical/Vet	
Other:	
______________________	
______________________	
______________________	
______________________	
______________________	

INCOME

Sales	
Other	

TOTAL EXPENDITURES ______________________

TOTAL INCOME ______________________

PROFIT / (LOSS) ______________________

PROFIT/LOSS STATEMENT

for January 1, 20_____ to December 31, 20_____

EXPENDITURES

Feed	
Supplies	
Equipment	
Maintenance/Repair	
Medical/Vet	
Other:	
____________	
____________	
____________	
____________	
____________	

INCOME

Sales	
Other	

TOTAL EXPENDITURES ____________________

TOTAL INCOME ____________________

PROFIT / (LOSS) ____________________

CONTACT LIST					
COMPANY	CONTACT PERSON	ADDRESS	PHONE #	EMAIL	WEBSITE

[illegible]

COMPANY	CONTACT PE[illegible]	[illegible]	[illegible]NE #	EMAIL	WEBSITE

CONTACT LIST

COMPANY	CONTACT PERSON	ADDRESS	PHONE #	EMAIL	WEBSITE

CALCULATION WORKSHEETS

CALCULATION SAMPLES

FEED COST CALCULATION

PRICE	BAG WGT	COST PER/LB
A	B	A ÷ B = C
$ 16.99	40	$ 0.42

AMOUNT OF FEED PER BIRD

# OF BIRDS	WGT OF FEED PER DAY	FEED/BIRD
D	E	D × E = F
25	3.5	0.14

DAILY FEED COST PER BIRD

FEED COST	WGT FEED/BIRD	DAILY COST
C	F	C × F = G
$ 0.42	0.14	$ 0.06

ROOSTER TO HEN RATIO

# HENS PER CAGE	# ROOS PER CAGE	TOTAL # OF BIRDS	BIRD RATIO
H	I	J	J ÷ H = K
4	1	5	1.25

AVG # DAYS PER EGG

AVG EGGS PER DAY	AVG # DAYS PER EGG
L	1 ÷ L = M
0.75	1.33

COST PER EGG

DAILY COST PER BIRD	BIRD RATIO	FEED COST OF BREEDING GRP	AVG # DAYS PER EGG	COST PER EGG
G	K	G × K = O	M	O × M = N
$ 0.06	1.25	$ 0.07	1.33	$ 0.10

COST PER MEAT BIRD

DAYS TO HARVEST	DAILY COST PER BIRD	COST PER BIRD
P	G	P × G = Q
56	$ 0.06	$ 0.59

COST PER LIVE BIRD SOLD

DAYS TO SALE	DAILY COST PER BIRD	COST PER BIRD
R	G	R × G = S
10	$ 0.06	$ 0.59
20	$ 0.06	$ 1.19
30	$ 0.06	$ 1.78

FEED COST CALCULATIONS

PRICE	BAG WGT	COST PER/LB
A	B	A ÷ B = C
$		$
$		$
$		$
$		$
$		$
$		$
$		$
$		$
$		$
$		$
$		$
$		$
$		$
$		$
$		$
$		$
$		$
$		$
$		$
$		$
$		$
$		$
$		$
$		$
$		$
$		$
$		$

PRICE	BAG WGT	COST PER/LB
A	B	A ÷ B = C
$		$
$		$
$		$
$		$
$		$
$		$
$		$
$		$
$		$
$		$
$		$
$		$
$		$
$		$
$		$
$		$
$		$
$		$
$		$
$		$
$		$
$		$
$		$
$		$
$		$
$		$
$		$

A… F FEE… BIR… ULATIONS

# OF BIRDS	WEIGHT OF FEED/DAY	FEED/BIRD
D	E	D × E = F

FEED COST PER BIRD CALCULATIONS

FEED COST PER LB.	WEIGHT OF FEED/BIRD	DAILY COST
C	F	$C \times F = G$
$		$
$		$
$		$
$		$
$		$
$		$
$		$
$		$
$		$
$		$
$		$
$		$
$		$
$		$
$		$
$		$
$		$
$		$
$		$
$		$
$		$
$		$
$		$
$		$

FEED COST PER LB.	WEIGHT OF FEED/BIRD	DAILY COST
C	F	$C \times F = G$
$		$
$		$
$		$
$		$
$		$
$		$
$		$
$		$
$		$
$		$
$		$
$		$
$		$
$		$
$		$
$		$
$		$
$		$
$		$
$		$
$		$
$		$
$		$
$		$

ROOSTER TO HEN RATIO

# HENS PER CAGE	# ROOS PER CAGE	TOTAL # OF BIRDS	BIRD RATIO
H	I	J	J ÷ H = K

# HENS PER CAGE	# ROOS PER CAGE	TOTAL # OF BIRDS	BIRD RATIO
H	I	J	J ÷ H = K

COST PER EGG

DAILY COST PER BIRD	BIRD RATIO	FEED COST OF BREEDING GRP	AVG # DAYS PER EGG	COST PER EGG
G	K	G × K = O	M	O × M = N
$		$		$
$		$		$
$		$		$
$		$		$
$		$		$
$		$		$
$		$		$
$		$		$
$		$		$
$		$		$
$		$		$
$		$		$
$		$		$
$		$		$
$		$		$
$		$		$
$		$		$
$		$		$
$		$		$
$		$		$
$		$		$
$		$		$
$		$		$
$		$		$
$		$		$
$		$		$
$		$		$
$		$		$
$		$		$
$		$		$

DAILY COST PER BIRD	BIRD RATIO	FEED COST OF BREEDING GRP	AVG # DAYS PER EGG	COST PER EGG
G	K	G × K = O	M	O × M = N
$		$		$
$		$		$
$		$		$
$		$		$
$		$		$
$		$		$
$		$		$
$		$		$
$		$		$
$		$		$
$		$		$
$		$		$
$		$		$
$		$		$
$		$		$
$		$		$
$		$		$
$		$		$
$		$		$
$		$		$
$		$		$
$		$		$
$		$		$
$		$		$
$		$		$
$		$		$
$		$		$
$		$		$
$		$		$
$		$		$

COST PER BIRD

DAYS TO HARVEST	DAILY COST PER BIRD	COST PER BIRD
P	G	P × G = Q
	$	
	$	
	$	
	$	
	$	
	$	
	$	
	$	
	$	
	$	
	$	
	$	
	$	
	$	
	$	
	$	
	$	
	$	
	$	
	$	
	$	
	$	
	$	
	$	
	$	
	$	
	$	

DAILY COST PER BIRD	COST PER BIRD
G	P × G = Q
$	$
$	$
$	$
$	$
$	$
$	$
$	$
$	$
$	$
$	$
$	$
$	$
$	$
$	$
$	$
$	$
$	$
$	$
$	$
$	$
$	$
$	$
$	$
$	$
$	$
$	$
$	$

COST PER LIVE BIRD SOLD

DAYS TO SALE	DAILY COST PER BIRD	COST PER BIRD
R	G	$R \times G = S$
	$	$
	$	$
	$	$
	$	$
	$	$
	$	$
	$	$
	$	$
	$	$
	$	$
	$	$
	$	$
	$	$
	$	$
	$	$
	$	$
	$	$
	$	$
	$	$
	$	$
	$	$
	$	$
	$	$
	$	$
	$	$
	$	$
	$	$

DAYS TO SALE	DAILY COST PER BIRD	COST PER BIRD
R	G	$R \times G = S$
	$	$
	$	$
	$	$
	$	$
	$	$
	$	$
	$	$
	$	$
	$	$
	$	$
	$	$
	$	$
	$	$
	$	$
	$	$
	$	$
	$	$
	$	$
	$	$
	$	$
	$	$
	$	$
	$	$
	$	$
	$	$
	$	$
	$	$

NOTES

NOTES

NOTES

NOTES

Made in the USA
Monee, IL
05 November 2026

67374700R00083